CRANBERRIES

by Golriz Golkar

Cody Koala

An Imprint of Pop!
popbooksonline.com

abdobooks.com
Published by Pop!, a division of ABDO, PO Box 398166, Minneapolis, Minnesota 55439.

Printed in the United States of America, North Mankato, Minnesota.

052021
092021

THIS BOOK CONTAINS RECYCLED MATERIALS

Cover Photos: Shutterstock Images, foreground; Michael P. Gadomski/Science Source, background
Interior Photos: Shutterstock Images, 1 (foreground), 8; Michael P. Gadomski/Science Source, 1 (background), 5; Red Line Editorial, 7 (top); Gregory K. Scott/Science Source, 7 (bottom left); iStockphoto, 7 (bottom right), 10, 14–15, 16, 17, 19 (top), 19 (bottom left), 19 (bottom right), 20; Tierbild Okapia/Science Source, 13

Editor: Aubrey Zalewski
Series Designers: Laura Graphenteen and Colleen McLaren

Library of Congress Control Number: 2020948888
Publisher's Cataloging-in-Publication Data
Names: Golkar, Golriz, author.
Title: Cranberries / by Golriz Golkar
Description: Minneapolis, Minnesota : Pop!, 2022 | Series: How foods grow | Includes online resources and index.
Identifiers: ISBN 9781532169809 (lib. bdg.) | ISBN 9781098240738 (ebook)
Subjects: LCSH: Cranberries--Juvenile literature. | Cranberry industry--Juvenile literature. | Bog plants--Juvenile literature. | Agriculture--Juvenile literature. | Food crops--Juvenile literature.
Classification: DDC 631.5--dc23

Hello! My name is

Cody Koala

Pop open this book and you'll find QR codes like this one, loaded with information, so you can learn even more!

Scan this code* and others like it while you read, or visit the website below to make this book pop.

popbooksonline.com/cranberries

*Scanning QR codes requires a web-enabled smart device with a QR code reader app and a camera.

Table of Contents

Chapter 1

What Are Cranberries?

Cranberries are small, round fruits. They can be red, pink, or white. They taste **tart**. They grow on low vines. The vines need lots of water to grow.

Watch a video here!

Chapter 2

Growing Cranberries

Cranberry plants grow in Canada and the northern United States. They grow in **bogs**. Bogs are near **wetlands**. Bogs have sandy soil.

Where Cranberries Grow

1. British Columbia, Canada
2. Quebec, Canada
3. Massachusetts, Unites States
4. New Jersey, United States
5. Wisconsin, United States
6. Washington, United States
7. Oregon, United States

Complete an activity here!

In winter, growers flood the bogs. The top of the water freezes. The ice layer and the water below protect the plants from winter weather.

Cranberries are mostly made of water.

Growers spread sand over the ice every two to five years. The sand sinks when the ice melts. It mixes with the soil. The sand is good for the cranberry plants. In the spring, growers pump out the water.

The plants flower in June. Their flowers are pinkish white and bell-shaped. The fruit grows between July and October.

Chapter 3

Harvesting Cranberries

In the fall, ripe cranberries are ready to **harvest**. The harvest lasts from September to November. Growers can harvest millions of pounds of cranberries.

Learn more here!

Growers flood the **bogs** again. Large machines stir the water.

Cranberries have four air pockets. These pockets help them float on water.

This makes the cranberries fall off their vines. Growers then sweep up the floating cranberries.

Chapter 4

Cranberry Foods

Cranberry sauce is a popular food made from cranberries. Americans often eat it at Thanksgiving. People also drink cranberry juice. Sugar and other fruits sweeten it.

Learn more here!

People also eat dried and sweetened cranberries. Dried cranberries are often an **ingredient** in snacks and baked treats.

Making Connections

Text-to-Self

Cranberries are tart. Do you like tart foods? Why or why not?

Text-to-Text

Have you read about another kind of fruit? How does it grow differently than cranberries?

Text-to-World

Cranberries grow in the United States and Canada. What is another crop that grows in the United States or Canada?

Glossary

bog – wet, spongy ground.

harvest – to gather or pick crops.

ingredient – one substance used in a mixture.

tart – sharp or sour to the taste.

wetland – an area often covered with water.

Index

Online Resources

popbooksonline.com

Thanks for reading this Cody Koala book!

Scan this code* and others like it in this book, or visit the website below to make this book pop!

popbooksonline.com/cranberries

*Scanning QR codes requires a web-enabled smart device with a QR code reader app and a camera.